A key to the crabs and crab-like animals of British inshore waters

By John Crothers and Marilyn Crothers

FSC

BRINGING
ENVIRONMENTAL
UNDERSTANDING TO ALL

First Edition 1983. Reprinted with minor alterations 1988. Reprinted 2014.
© FSC 1983, 2014.
ISBN 978 1 85153 155 4
Publication code 155

Acknowledgements

Very many people have offered us advice and help in the preparation of this key. We are grateful to all, but especially so to Drs R. W. Ingle and R. G. Hartnoll for putting us right on factual matters, and to Claire Dalby for advice on methods of presentation.

Comments on the Test Version were received from 122 individuals or groups of people during the two seasons for which it was available for evaluation. The testers included sixth form and university students, teachers and university lecturers, both specialist and non-specialist research workers and people involved in survey work.

We are most grateful for these comments, and for the work of Steve Tilling (AIDGAP Co-ordinator) who collated them and produced a six page report on the results of the testing programme.

There can be no doubt but that this published version is a great deal more useful as a result of the AIDGAP system.

This AIDGAP key was first published in *Field Studies* **5**: 753-806 and was reprinted with minor alterations in 1988. It has reprinted with a new cover and revised page numbering in 2014.

An amendment to couplet 21 was made in 1990 (*Field Studies* **7**: 483-484). In the 2014 reprint, the 1990 amendment has been fully integrated into the main body of the key.

Contents

Introduction

It is hoped that this key will enable anybody to identify all the 'crabs' they may find on British sea shores and whilst skin-diving or dredging close inshore. There should be no need to kill the animal, or to extract it from its shell (if a hermit crab) in order to find out what it is.

This is an AIDGAP publication (Aids to Identification in Difficult Groups of Animals and Plants). The Difficulty in this case lies in the problems encountered by a non-specialist when trying to use any of the existing keys, either because of their limited scope (keys to one genus with no guide as to how that genus is separated from others), because of the obscure characters employed, and/or because of their incompleteness.

The present authors have thus used non-technical language wherever possible, and have illustrated all the terms employed in the explanatory figures. There may be difficulties in seeing some of the characters on very small specimens, but almost all should be distinguished if a x10 hand lens is used. Each species is fully illustrated on pages 35-57. These drawings should be used to CHECK the identity of the crab identified from the text. For reasons that are explained later, each species is very variable in size and shape: it is not possible to identify all crabs from pictures alone, and not all individuals of a species will look like the drawings.

The decision as to which species to include and which to ignore was, of necessity, subjective. On the one hand it was decided to include all the animals generally called 'crabs' by non-specialists – so hermit and porcelain crabs are here but not shrimps, prawns, lobsters or squat lobsters. "Inshore Waters" is taken to mean the sea bed down to about 30 m within about 2 km of the land. With increasing depth and distance offshore other species may be encountered. The key is annotated to allow for their existence and the reader is recommended to turn to Ingle (1980) for further information. "British Waters" means all the British Isles, including Ireland and the Channel Islands, but in fact this key will probably work satisfactorily on the adjacent continental coasts as well – except for some freshwater species now occurring in Dutch and German estuaries.

The present authors have carried out no original taxonomic work. We have neither lumped nor split any species, and have utilised hardly any new characters in compiling the key. What we have done is to select those features that are most readily seen in the living animal and to illustrate the species clearly. It is not a "natural" key. It makes no attempt to key out families, then genera, and finally species (unless this happens to be convenient). Such keys have their place. In some ways they justify the classification system. But this key is written for the non-specialist who wishes to put a name to the crab in his hand, and has only a passing interest in taxonomy.

The specimens were identified from Allen (1967), Bell (1853), Christiansen (1969), Bouvier (1940), Forest (1978) for *Macropodia*, Hartnoll (1963) for other spider crabs, Leach (1814-1875), Palmer (1927) for *Liocarcinus* and *Macropipus*, and Selbie (1921) for hermit crabs. Most of the illustrations are of specimens in th authors' collection. Other species were kindly made available to us by Dr. R. W Ingle of the British Museum (Natural History). Without his assistance and permission to examine the museum material this work would not have been possible. Reference should also be made to Dr. Ingle's (1983) Linnean Society synopsis on shallow water crabs and to his (1985) revision of *Pagurus*.

The result is a compilation of many people's work, adapted and re-written for the non-specialist in the light of the authors' experience. The first version was written for students attending field courses at Dale Fort Field Centre (Dyfed) in about 1964/65 and a modified text appeared in the Field Studies Council's internal bulletin "Fieldwork". As a result of the comments received a third version was circulated around certain marine Field Centres in 1976/77. An AIDGAP test version, in the present format, was produced in 1979/80 and subjected to extensive user-testing during 1981 and 1982. The present edition has been revised to incorporate many of the comments and suggestions thrown up by the testing process.

Basic crab biology

It is necessary to know a little about crab anatomy in order to use this key, and may be helpful to know something of their method of growth as well. A more extensive treatment will be found in Crothers (1967:1968). See Lancaster (1988) for hermits.

Anatomy

Crabs are crustaceans and, like lobsters, prawns, and woodlice, they have a hard exoskeleton (a skeleton on the outside of the body) and numerous jointed limbs. In the woodlouse it is clear that the limbs are segmentally arranged. Each body segment has a pair of legs, and all the legs are of similar size and shape. In the more advanced crustacea, such as crabs, there has been considerable specialisation of function with the result that no two pairs of legs are exactly alike. Table 1 lists the appendages of a typical crab, and the conspicuous ones are labelled in Figs 1 and 2.

When seen from underneath, the segmental arrangement of the walking legs is clearly visible, but from above there is no trace of its existence. This is because the dorsal surface of the body is hidden by a fold of skin, the carapace, which has grown sideways from the mid-line, out and over the gills, enclosing the latter in gill chambers.

In general males grow larger than females, and have longer legs, bigger chelae and a more aggressive manner. The male abdomen is usually narrower than that of the female, and often appears to have fewer segments.

The hard rigid exoskeleton of the crab represents up to 40 per cent of its weight and not only protects the animal from damage but also provides anchorage for muscle attachment. It is jointed in places to permit movement. The hard sections, impregnated with calcium salts, hinge on each other by ball-and-socket joints and are linked by pliable sections. This rigid skeleton is non-living and, once hardened, it cannot change shape. The crab can only grow by moulting.

Moulting

The animal must lay down a new skeleton within the old, resorbing and redistributing as much material as possible. The new skin is corrugated, so as to fit within the confines of the old shell. The crab then finds somewhere safe to hide and cracks off the old shell by drinking a large quantity of water. Splits appear down the sides of the carapace, along a pre-formed line of weakness, and

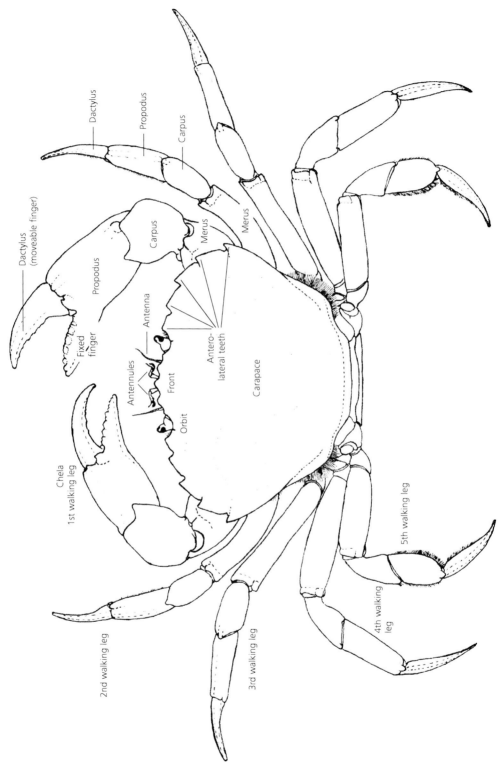

Figure 1. A Brachyrhynchan crab (*Carcinus maenas*) labelled to indicate some of the characters used in the key. Note that the term "front" is applied to the area of the carapace between the eyes. In species where the front projects forward, well beyond the eyestalks, it is the "rostrum".

Table 1. The segmental nature and function of the appendages in *Carcinus maenas*, and most other species, after Crothers (1967).

Segment	Appendages	Tagma	Function
1	nil	Head	(Embryonic)
2	Antennules		Sensory: The statocyst is in the basal section – keeps crab right way up
3	Antennae		Sensory
4	Mandibles		Cut up food
5	Maxillules		Hold food for mandibles to cut
6	Maxillae		Pass food to mandibles. An expodite (the scaphognathite) produces the respiratory current through the gill chamber
7	1st Maxillipeds	Cephalothorax	Pass food to maxillae
8	2nd Maxillipeds		Pass food to maxillae
9	3rd Maxillipeds		Grasp food, manipulate it and pass it forward to the other mouth parts
10	Chelae (1st Walking Legs)	Thorax	Offence/defence: pick up/tear off pieces of food and pass them to maxillipeds
11	2nd Walking Legs		Walking: sensory at tip
12	3rd Walking Legs		Walking: sensory at tip
13	4th Walking Legs		Walking: sensory at tip
14	5th Walking Legs		Walking/burrowing/swimming

Segment	Appendages	Tagma	Male	Female
15	1st Pleopods	Abdomen	Copulatory styles	Absent
16	2nd Pleopods		Copulatory styles	Used for carrying eggs
17	3rd Pleopods		Absent	Used for carrying eggs
18	4th Pleopods		Absent	Used for carrying eggs
19	5th Pleopods		Absent	Used for carrying eggs

gradually work backwards to meet across the rear end of the carapace. The animal then carefully withdraws its body from the old shell and goes on drinking water to stretch the new one to its required size. In *Carcinus* a 30 per cent increase in size at each moult is commonplace. The new skeleton is soft and easily damaged (beware of damaged specimens when using this key). Not only is the crab vulnerable to predators at this time, but also it has difficulty in moving about as the muscles simply bend the shell. *Carcinus* may take up to two or three hours to cast its old skeleton, and then require anything from three days to a fortnight to harden the new one. The time depends on the size of the crab and the temperature: the larger the crab and the colder the water the longer it takes.

So soft crabs are not a separate species: they simply represent short interludes in the lives of all individuals.

I think *Carcinus* can expect to moult about 18 times during its four years of life – mostly during the first year – and imagine that several other species are similar. Larger and longer-lived species such as *Cancer* presumably moult even more often.

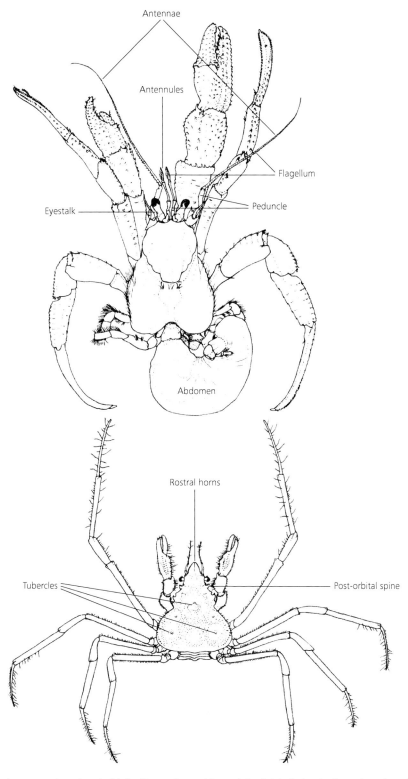

Figure 2. (a) An Anomuran hermit crab (b) An Oxyrynchan spider crab both labelled to indicate the other characters used in the key. Names for the parts of the legs and so on are the same as in Fig. 1.

Growth

Growth is allometric, that is, some parts grow faster than others so that the overall form gradually changes proportion with age. Young *Cancer*, for example, are nearly circular whilst adults are definitely oval.

Should a crab be so unfortunate as to lose a leg, it is able to regenerate it at the next moult. The new appendage will be much smaller than the original but grows disproportionately fast so that it reaches full size within four or five moults. Most pairs of legs are similar, that is the right is the same as the left. But as regards the chelae, one is almost always bigger than the other. Most individuals seem to be right handed – the right chela is bigger than the left. If this master chela is lost and has to be regenerated, the left chela adopts the form of the damaged right at the next moult.

Problems in identification

In several species, it is the young crabs that are found on the shore, and the older ones stay in deeper water. As in many other groups, young individuals are easier to confuse with others of similar species than are their parents.

Then there are the differences between males and females: both sexes grow allometrically, and are often missing limbs. It should be possible to identify damaged specimens from this key, but users are recommended to try with complete ones first. When in doubt consult Ingle (1980).

How to use the key

With your unknown crab beside you, read the first couplet, which asks whether your specimen is a hermit crab (protecting its abdomen in a snail shell) or a normal sort of crab. If you think yours is a hermit you proceed to couplet 6. Couplet 6 asks you whether the left chela is bigger than the right. If it is, your specimen is probably *Diogenes pugilator* but, remembering that the left chela might be bigger simply because the crab lost the right one some time ago, read the rest of the description to check that it fits and then have a look at the illustrations. Keys are really very simple if you proceed in a logical manner and do not try to take short cuts.

The key

1. Hermit crabs, usually protecting their soft abdomen with a snail shell. Those that have lost their shells carry their abdomen curled up behind them 6

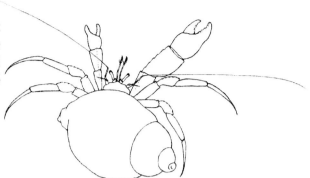

- Crabs, or crab-like animals, not living in empty snail shells (although may be within the shell of a living bivalve). The abdomen is carried tucked underneath the body 2

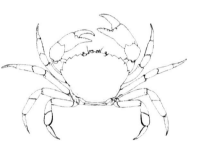

2. Crabs with four pairs of walking legs in addition to the chelae (claws) [beware mutilated specimens] 15

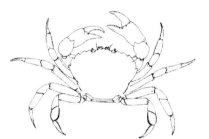

- Crab-like animals with apparently only three pairs of walking legs in addition to the chelae. There may be a much-reduced fifth pair tucked up inside the abdomen, or around the sides of the carapace 3

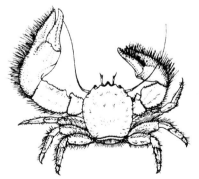

3. Chelae shorter than the other walking legs.
   Body and legs bear conspicuous sharp spines.
   Carapace up to about 150 mm long. Mainly
   north and east coasts ....
   .... **The Stone Crab** *Lithodes maia*

- Chelae larger and longer than the other walking
  legs 4

4. Between the eyes the carapace projects markedly forwards
   to form a rostrum bearing conspicuous sharp spines. Squat
   Lobsters, not considered further in this key.

- Carapace does not project forward to form a rostrum 5

5. Body and chelae hairy. Chelae very large and flattened. May
   be found on the shore ....
   The Broad-clawed Porcelain Crab *Porcellana platycheles*

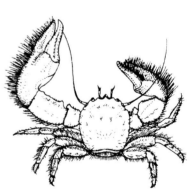

- Body and chelae smooth. Chelae very long and cylindrical.
  May be found on the shore ....
  .... **The Long-clawed Porcelain Crab** *Pisidia longicornis*

6. The left chela much larger than the right, and fairly smooth. The right chela densely covered with hairs. In life the animal is bluish in colour. This is especially noticeable on the chelae and around the head region. Carapace up to about 10 mm long. Southwestern coasts, may be found on the shore .... *Diogenes pugilator*

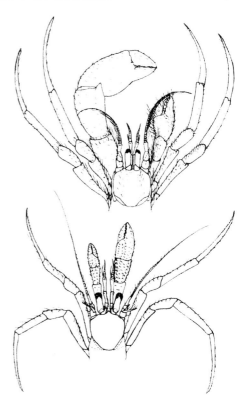

- Right chela larger than the left 7

7. The snail shell inhabited by the hermit has an anemone or a hydroid growing on it 8

- Shell without anemone or hydroid colony 9

8. Anenome *Adamsia maculata* (= *A. carciniopados* and *A. palliata*): the tentacles sweep the ground under the shell. The disc frequently envelopes the shell so that its two sides meet down the dorsal mid-line. The hermit crab has rather smooth chelae, and large black eyes. There is a red area on the eyestalk just behind the eye (see also couplet 14) .... *Pagurus prideaux*

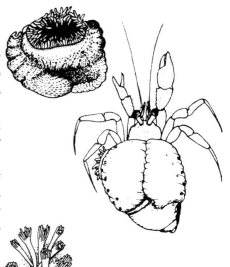

- Hydroid colony *Hydractinia echinata* or anemone *Calliactis parasitica*: the tentacles face upwards and the disc is attached to the top of the shell. Hermit crab is very probably .... *Pagurus bernhardus* (See also couplet 13)

(Left) a tower shell, formerly inhabited by *Pagurus bernhardus* covered in the hydroid *Hydractinia echinata*. (Right) an enlargement of the hydroid polyps.

9. Right chela with conspicuous long or short hairs 10

- Right chela practically hairless 12

10. All visible parts of the crab are very hairy in life. The hairs on the right chela form a matted fur. The long eyestalks nearly reach to the end of the antennular peduncle, and are about twice as long as its basal segment.

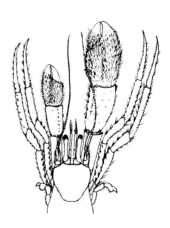

The front of the carapace between the eyes is rounded. Carpus and propodus of the right chela are of about equal length*. Carapace up to about 15 mm long. May be found on the shore .... *Pagurus cuanensis*

- The hairs on the right chela do not form a matted fur. The eyestalks are about three-quarters the length of the antennular peduncle 11

11. The hairs on the right chela are short and grow in clumps from the base of tubercles. The carpus of the right chela is nearly as long as the propodus and slightly wider*. The front of the carapace between the eyes has a sharp point. Carapace up to about 30 mm long. Not found on the shore .... *Pagurus pubescens*

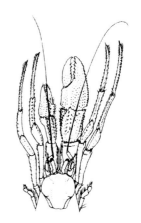

- The hairs on the right chela are longer, arise singly and are generally distributed over the surface. The propodus of the right chela is usually longer and much wider than the carpus*. The front of the carapace between the eyes rounded. Carapace up to about 5.5 mm long. Not found on the shore .... *Anapagurus chiroacanthus*

*The length of the propodus is taken to include the fixed finger.

12. The propodus of the right chela is appreciably wider and shorter than the carpus, often white in colour. Antennular peduncle long (more than twice the length of the eyestalk) and fringed with long hairs on the inside. Carapace up to about 7 mm long. May be found on the shore ....
.... *Anapagurus hyndmanni*

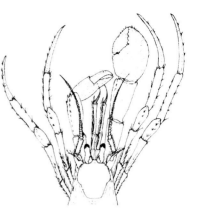

- The propodus of the right chela is little wider and usually much longer than the carpus. Antennular peduncle is not twice as long as the eyestalks 13

13. The propodus of the right chela is yellowish in colour, usually with a conspicuous red line down the middle. The surface of the right chela is covered with tubercles, is gently convex, and there is a row of slightly larger ones on either side of the red line. There is a row of slender transparent spines along the lower edge of the dactylus on each leg (only visible on larger specimens except under a microscope). Carapace up to about 35 mm long. Smaller ones are often very common on the shore ....
.... *Pagurus bernhardus*

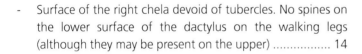

- Surface of the right chela devoid of tubercles. No spines on the lower surface of the dactylus on the walking legs (although they may be present on the upper) 14

14. There is (in life) a red area on the eyestalks just behind the eye. There is a row of low blunt teeth around the margin of the fixed finger on the right chela and another row, perhaps of spines, down the mid line of the carpus. Carapace up to about 30 mm long. Not usually found on the shore ....
.... *Pagurus prideaux*

Living specimens almost always have the anemone *Adamsia maculata* (= *A. carciniopados* and *A. palliata*) on their shell. See Manuel (1981), Erwin and Picton (1987).

- No red area on the eyestalk. There are no teeth along the edge of the fixed finger of the right chela and any spines tend to be along the margins, not the centre, of the carpus. Carapace up to about 8 mm long. Not usually found on the shore .... *Anapagurus laevis*

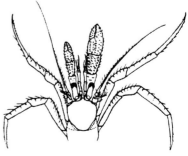

15. The fourth and fifth pairs of walking legs are smaller than the others and are chelate (i.e. form nippers). In life they are used to carry a piece of sponge or other camouflage over the carapace. The whole body covered in a felt of dark brown hairs. The tips of the fingers on the, very strong, chelae are rose pink in colour, like nail varnish. Carapace up to 80 mm or more across. Southern coasts. Not usually found on the shore .... **The Sponge Crab** *Dromia personata*

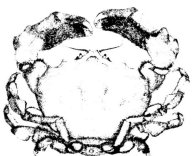

-    Only the first pair of walking legs chelate.

16. Between the eyes are two prominent rostral horns (although in some species these are separated by a very narrow cleft). Antennae are never densely fringed with hairs. Carapace pear-shaped or roughly triangular with the apex pointing forwards. Some species have a spidery appearance 43

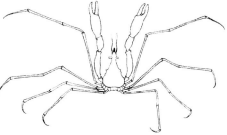

-    No rostral horns. Only one species has two teeth on the front of the carapace between the eyes and that has long antennae fringed with hairs. Carapace is usually broader than long. If roughly triangular then the apex points backwards. None of these species have a spidery appearance 17

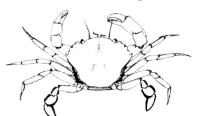

17. Small crabs, carapace up to about 7 mm across, found within the shelll of a living bivalve, usually a mussel, cockle or oyster. Females are larger than males. In both sexes the eyes and orbits (eye-sockets) are very small and the rounded carapace bears no marginal teeth .... **Pea Crabs** ... 18

-    Free-living crabs 19

18. The dactylus of each walking leg is about half the length of the propodus with a pronounced inward curve at the end. The front of the carapace is smooth or curved outwards. Usually found in mussels, *Mytilus* or cockles, *Cerastoderma*, May be found on the shore ... ***Pinnotheres pisum***

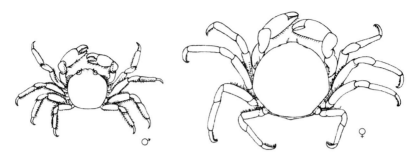

- The dactylus of each walking leg is appreciably more than half the length of the propodus and has only a slight inward curve. The front of the carapace has a slight median notch, more obvious in males. Male has very characteristic pattern of hairs on the walking legs: 2nd pair (i.e. those immediately behind the chelae) only bear hairs on the under side; in the 3rd and 4th pair the merus has hairs along the under side but on the carpus, propodus and dactylus the hairs arise from the upper face to hang down over the lower edge. The 5th pair has hairs on the upper surface of the merus and the lower surface of the other sections. Usually found in horse mussels, *Modiolus*, oysters, or fan mussels *Pinna*. Not usually found on the shore. Southern coasts .... ***Pinnotheres pinnotheres***

19. Antennae long and very hairy 20
    (look at the drawings for couplet 20)

- Antennae short and bearing only sparse hairs 21

20. Antennae shorter than the length of the carapace. Carapace roughly circular with 8-10 triangular teeth along both lateral margins. The sides of the carapace and legs bear long hairs. Chelae usually have black tips to the fingers. Carapace up to about 30 mm across. Not usually found on the shore ....
.... **The Circular Crab** *Atelecydus rotundatus*

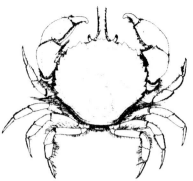

- Antennae longer than the carapace. When brought together the hairs on the antennae interlock to form a tube down which the crab can draw water when burrowed in the sand. Four sharp teeth on the outer margins of the carapace: two anterolaterals, one mediolateral and one (very small) posterolateral. Male has very long chelae but they never have black tips. Carapace longer than broad, about 40 mm in length. May be found on the shore, sandy beaches .... **The Masked Crab** *Corystes cassivelaunus*

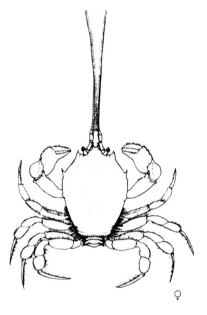

21. Distal segments (dactylus and propodus) of each walking leg with a row of spines along the underside. Carapace almost square and with very large eyes. Dactylus of the fifth (last) walking leg slightly flattened, but not expanded as a paddle. Carapace up to 20 mm across ...... *Planes minutus*

- Dactylus and propodus of each walking leg without spines, although they may be fringed or covered by hairs. If the carapace approximately square, then with small eyes. Carapace may be much more than 20 mm across 21a

21a. Dactylus of the fifth (rear) walking leg cylindrical in cross-section and always narrower than the propodus. It may have hairs on it, but they do not form a dense fringe around the border.

- Dactylus of the fifth (rear) walking leg flattened dorso-ventrally; often expanded as a paddle and usually fringed with dense hairs, at least along one border.

NOTE. In the shore crab the dactylus of the fifth walking leg is flattened but NOT expanded.

22. Dactylus of the fifth walking leg flattened and expanded as a paddle .... **Swimming Crabs**

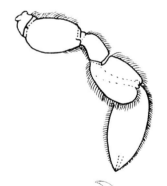

- Dactylus of the fifth walking leg flattened but not expanded. Small specimens have a fringe of hairs around the dactylus but these are often worn away in older ones. Carapace broader than long. Between the eyes, the front has three rounded lobes. There are five sharp anterolateral teeth, all much the same size. Colour variable, usually green or red, but small individuals are often patterned with symmetrical white, black or red markings. Carapace up to 88 mm across, but those on the shore usually up to about 50 mm. Common on the shore and down to about 5 m ......
.... **The Shore Crab** *Carcinus maenas*

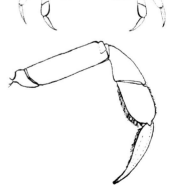

23. Carapace broader than long or roughly circular 24

- Carapace longer than broad although the discrepancy is not always so marked as in the specimen illustrated. Dactylus of the fifth walking leg pointed and bearing a fringe of hairs along the inside margin only. Sandy coloured with a pattern of small circles all over the carapace. Carapace up to about 20 mm across. May be found on the shore, sandy beaches .... *Portumnus latipes*

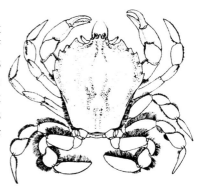

24. Front of the carapace with three or more teeth or lobes between the eyes 25

- Front of the carapace entire, perhaps with a fringe of hairs. The five anterolateral teeth on the carapace are markedly unequal, with the third and fifth larger than the others. Dactylus of the fifth walking leg sharply pointed and fringed with hairs on both sides. Carapace up to about 30 mm across: colour sandy. Not usually found on the shore .... *Liocarcinus arcuatus*

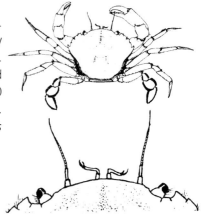

25. Front with three teeth or lobes between the eyes. Eyes dark 26

- Front with many (8-10) small teeth between the eyes. Eyes red. Legs bear blue lines on all sections (red after death), most noticeable on the swimming paddles. Carapace covered in a felt of dark brown hairs. Carapace up to about 90 mm across. May be found on the shore .... **The Velvet Swimming Crab** *Necora puber*

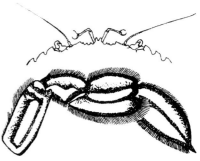

26. Only the fifth walking leg obviously adapted for swimming 26

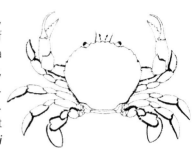

- Dactylus and propodus of all legs flattened for swimming, and fringed with hairs, at least along the inside. Dactylus of the fifth leg very much expanded and the propodus has a large, rounded, distal lobe. Large eyes. Carapace circular, frontal teeth often more prominent than the anterolaterals. Carapace up to about 45 mm across. A southern species. Not usually found on the shore, but may swim over it at high tide .... *Polybius henslowii*

Distal lobe of the propodus ⎯⎯⎯

27. Front of the carapace projects forward between the eyes, bearing three rounded lobes. Propodus of the fifth walking leg does not have a distal lobe. Surface of the carapace rough. Carapace up to about 25 mm across. May be found on the shore .... *Liocarcinus pusillus*

[*L. zariquieyi* is very similar. But the central lobe of the front does not project beyond the outer ones. Outside anterolateral tooth much reduced. Habitat unknown; see Ingle (1980:1983)]

[The southern species *Xaiva biguttata* = *Portumnus biguttata* would also key out here. The dactylus of the fifth walking leg bears a fringe of hairs along the inside only. Carapace surface smooth. Small eyes. Not usually found on the shore]

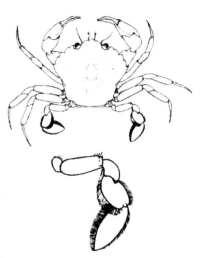

- Front of the carapace projects very slightly if at all but bears three teeth between the eyes 28

28. Dactylus of the fifth walking leg with a median rib (appears as a raised ridge) 29

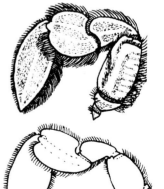

- Dactylus of the fifth walking leg without a median rib 30

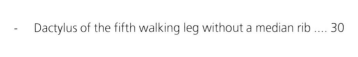

29. Anterolateral teeth on the carapace are all approximately equal in size. Surface of the carapace covered in transverse rows of crenulations, each set with a fringe of hairs. Colour ranges from light brown to red. Carapace up to about 40 mm across. Not found on the shore ....
.... *Liocarcinus corrugatus*

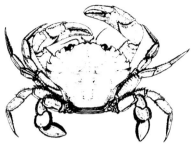

- Outside anterolateral teeth much longer than the others. Not found on the shore and usually in deeper water. Carapace up to about 60 mm ....
.... *Macropipus tuberculatus*

[*Bathynectes* species would also key out here. In these deep water crabs the outside anterolaterals are even more pronounced and joined together by a prominent ridge across the carapace.]

30. The rounded distal lobe of the propodus on the fifth walking leg has a 'straight end'. In large, living crabs the dactylus on the fifth walking leg is tinged with violet. This is very obvious in adults but much less so in young ones. Carapace up to about 40 mm across. May be found on the shore .... *Liocarcinus depurator*

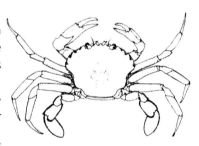

'Straight end' to the distal lobe of the propodus ——————

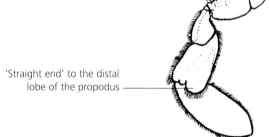

- Any distal lobe to the propodus of the fifth walking leg has a rounded end. The dactylus of the fifth walking leg is not tinged with violet 31

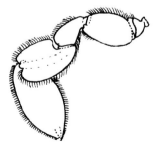

These two species are easily confused, especially when small.

31. The merus of the fourth walking leg is twice as long as the merus of the fifth. On the chela there are two teeth (lumps really) on the outside of the carpus where it joins the propodus. Between the eyes the front of the carapace bears three sharp teeth: the central one usually projects slightly forward of the other two. Colour often greenish. Carapace up to about 40 mm across. Does not usually occur on the shore .... *Liocarcinus holsatus*

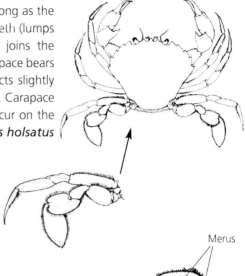

Merus

- The merus of the fourth walking leg is only 1.5 times the length of the merus of the fifth. On the chela, the outside margin of the carpus is smooth where it joins the propodus. Between the eyes, the front of the carapace bears three rounded teeth: the central one does not project. Colour patterned with various shades of brown. Carapace up to about 35 mm across. Does not usually occur on the shore .... **Marbled Swimming Crab** *Liocarcinus marmoreus*

Read three alternatives

32. Carapace with prominent marginal teeth, or lobes 33

- Carapace without marginal teeth 35

- Carapace with five very indistinct antero-lateral teeth, no teeth between the eyes, and with a prominent fringe of hairs. A southern species not usually found on the shore .... *Thia scutellata*

33. Carapace rectangular, much broader than long, with two lateral teeth. Very long eyestalks. The male has very long chelae; in the female they are of similar length to the other walking legs; the moveable finger is usually dark-coloured along the inside edge. Carapace about 40 mm across. May be found on the shore but much commoner below the tide-marks .... **Goneplax rhomboides**

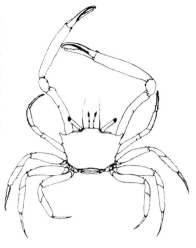

- Carapace with more than two lateral teeth or lobes. Eyestalks of normal length 34

34. From of the carapace projects slightly forward between the eyes and bears three sharp points. Five strong anterolateral teeth. May be found on the shore. Carapace about 20 mm across .... **Pirimela denticulata**

(Compare with *Carcinus* p. 15.)

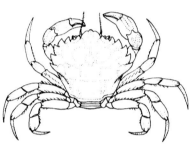

- Front of the carapace does not project forward between the eyes 39

35. Carapace oval. Chelae shorter than the other walking legs 38

- Carapace diamond-shaped. Chelae longer than the other walking legs. Antennae minute and the eyes very small. *Ebalia* sp. 36

36. Seen in lateral view, the posterior portion of the carapace is raised well above the anterior region. Seen from above, the tubercles on the carapace are fused together into a prominent cross. Abdominal segments 3-6 fused together in both sexes. Carapace about 18 mm across by 17 mm long. May be found on the shore .... **Ebalia tuberosa**

[The deeper water species *E. nux* would key out here on the first character but there is no sign of the cross and the carapace has pronounced bead-like tubercles. Not bigger than 9 mm across.]

- Seen in lateral view the silhouette is lumpy with no clear-cut difference in height between anterior and posterior regions. Seen from above, the tubercles are discrete. Males have abdominal segments 3-5 fused together, females 4-6 37

37. Seen in lateral view, the tubercles on the middle of the carapace, and at the rear are higher than the frontal region. Seen from above the carapace is slightly broader than long (13 mm across by 12 mm long), octagonal in outline. The surface of the carapace and the limbs is minutely granular. Not usually found on the shore .... *Ebalia tumefacta*

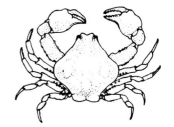

- Seen in lateral view the carapace is much flatter than in the other two species, rising anteriorly, in the middle (where there are two low tubercles either side of the mid line) and in a rounded bump at the posterior end. Seen from above the carapace is as long as broad (11 mm each way) and appears in outline like a diamond with the corners cut off. The surface of the carapace and limbs, especially the merus of the chelae, is obviously granular. Not usually found on the shore .... *Ebalia cranchii*

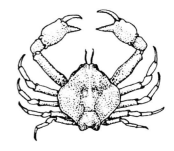

38. Lateral borders of the carapace completely smooth, and the surface without hairs or noticeable granules. **Pea Crabs** 18

- Carapace with an irregular margin or pie-crust edge. Surface may be very hairy. Tend to have large nippers 39

39. Carapace with a pie-crust edge, having about ten lobes on either side. Surface smooth and pink in colour. Black tips to the fingers. Legs coarsely hairy. Carapace up to about 160 mm across. Smaller individuals may be common on the shore, may vary greatly in colour (white and mottled cream/brown individuals are common) and are more nearly circular in shape .... **The Edible Crab** *Cancer pagurus*

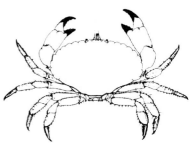

- Carapace without an even pie-crust edge 40

40. Carapace and legs covered in hairs, which are broader at their outer end than at the base. Legs obviously striped transversely. Beneath the hairs the carapace bears short sharp spines over the dorsal surface: these are easily seen in young crabs but become increasingly hidden by the hairs in older specimens. Five sharp anterolateral teeth. Fingers of the chelae brown at the tips. Males have one (usually the right) chela much larger than the other. Usually reddish in colour. Carapace about 20 mm across. May be found on the shore .... **The Hairy Crab** *Pilumnus hirtellus*

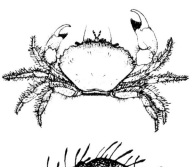

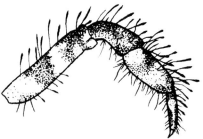

- Carapace not covered in hairs. Legs without transverse stripes 41

41. Walking legs with few hairs. Males much bigger than females, carapace up to about 60 mm across. May be found on the shore .... *Xantho incisus*

[Juvenile shore crabs *Carcinus maenas*, sometimes key out here. Return to couplet 22.]

- Walking legs bear conspicuous hairs on the dactylus and propodus 42

42. The hairs are confined to the dactylus and propodus. Carapace up to about 25 mm across. Not found on the shore and usually in deeper water ...... *Monodaeus couchi*

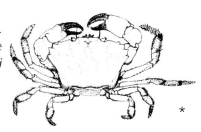

- A fringe of hairs runs along the whole length of the walking legs. Carapace up to about 35 mm (males larger than females). May be found on the shore. Fingers of the chelae may be darker than in the specimen illustrated .... *Xantho pilipes*

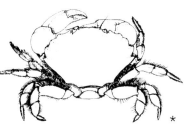

*Colour of the fingers is variable in these species. In *X. incisus* they are usually black and in *X. pilipes* brown but this is not diagnostic.

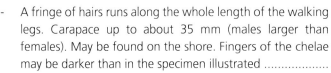

43. Walking legs much longer than the chelae. Spidery crabs. Orbits (eye-sockets) absent or represented by a single post-orbital spine. Abdomen of six segments 44

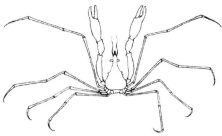

- Walking legs not much longer than the chelae, at least in the male. Not particularly spidery in appearance. Orbits are present, with a supra-orbital ridge anda post-orbital spine. The abdomen has seven segments 51

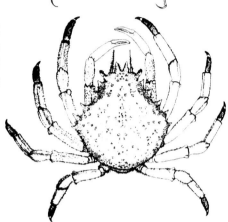

44. The eyestalks can be retracted against the post-orbital spines. The dactylus of the walking legs is straight or only slightly curved 45

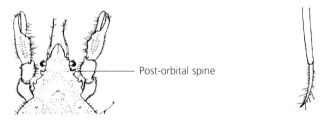

Post-orbital spine

- The eyestalks cannot be retracted. There are no post-orbital spines. The dactylus of the fourth and fifth walking legs is noticeably curved 47

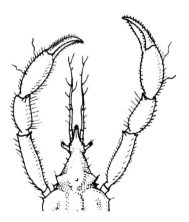

45. The rostral horns are separated by a broad, shallow U-shaped cleft. On the carapace there is a large central tubercle, with a row of four smaller ones in front of it. (It may be necessary to scrape sponges and other material off the crab in order to see these features.) Carapace up to about 30 mm but usually smaller. Not usually found on the shore ....
.... *Inachus dorsettensis*

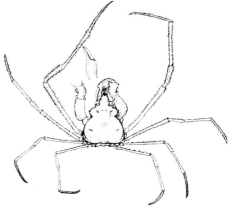

\- The rostral horns are separated by a narrow cleft. There are only two tubercles in the transverse row instead of four. (It may be necessary to scrape sponges and other material off the crab in order to see these features.) 46

46. The rostral horns are separated by a deep U-shaped slit, about twice as long as wide. There is a row of long hairs (about half the diameter of the cornea) along the anterior margin of the eyestalks. Adult males have a tubercle on the underside of the thorax, called the sternal callosity. Not usually found on the shore. Carapace up to about 28 mm long .... *Inachus leptochirus*

\- The rostral horns are separated by a very narrow slit. Any hairs on the eyestalk are less than a quarter the diameter of the cornea. Males do not have a sternal callosity. May be found on the shore. Carapace up to about 20 mm long ......
.... *Inachus phalangium*

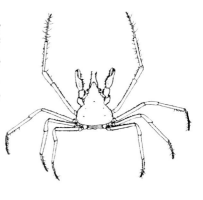

47. The rostrum is much longer than the antennal peduncle and is inclined upwards. The basal segment of the antennal peduncle bears a row of spines along its lower surface. Carapace up to about 30 mm long. Not usually found on the shore .... *Macropodia tenuirostris*

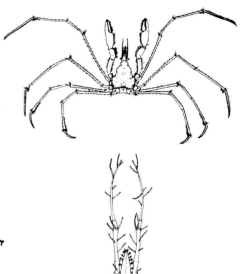

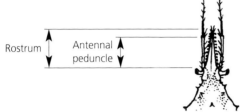

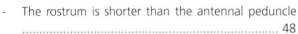

Rostrum   Antennal peduncle

- The rostrum is shorter than the antennal peduncle 48

48. The dactylus of the fourth and fifth walking legs is very strongly curved, almost semi-circular in shape with teeth along the inner margin. Rostrum very short; not much longer than its width across the base. Small crabs, carapace not much more than 15 mm long 49

- The dactylus of the fourth and fifth walking legs is not curved into a semi-circle and bears hairs, not spines along the inner margin. The rostrum is much longer than its basal width 50

49. The eyestalks have a distinct lump in the middle of their anterior margin. The rostral horns do not touch, and are shorter than the first segment of the antennal peduncle. Carapace up to about 10 mm long. Not usually found on the shore .... *Achaeus cranchii*

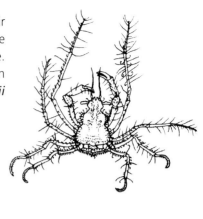

- The eyestalks have no lump but may show an indistinct spine in the middle of their anterior margin. Rostral horns are in contact for their entire length and extend well beyond the end of the first segment of the antennal peduncle. Carapace up to about 15 mm long. Not found on the shore .... *Macropodia linaresi*

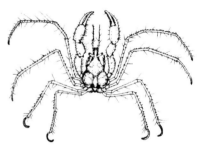

50. Rostrum about as long as the antennal peduncle, and usually with a distinct downward curve. The basal segment of the antennule peduncle may have several short spines. There may be a small lump on the anterior margin of the eyestalks. Carapace up to about 30 mm long. Not usually found on the shore .... *Macropodia deflexa*

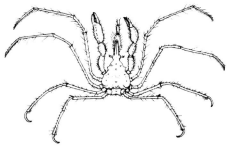

- Rostrum straight, and much less than half the length of the antennal peduncle. The basal segment of the antennule peduncle is smooth and devoid of spines in adults. No lump or spine on the anterior border of the eyestalks. Carapace up to about 25 mm long. May be found on the shore .... *Macropodia rostrata*

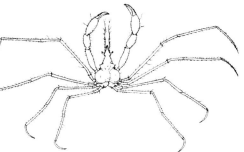

51. Carapace bearing raised mushroom-like tubercles. Rostral horns diverging and bearing a double row of hooked hairs along the inner edge of each horn .... *Eurynome* ... 52

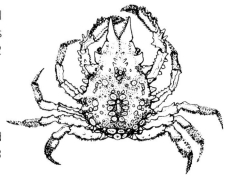

- Carapace may bear spines but not mushroom-shaped tubercles 53

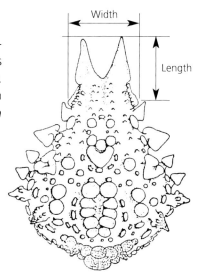

52. Basal width of the rostrum greater than the length. A U-shaped cleft between the rostral horns. The tubercles across the posterior edge of the carapace are fused into a row. Carapace up to about 17 mm long. Not usually found on the shore .... *Eurynome aspera*

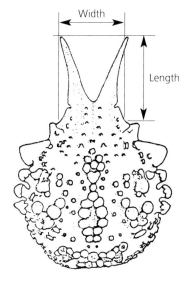

- Basal width of the rostrum less (males) or only slightly more (females) than the length. A V-shaped cleft between the horns. The tubercles across the posterior edge of the carapace are discrete or only slightly fused. Carapace up to 10.5 mm. May be found on the shore .... *Eurynome spinosa*

53. Carapace and legs completely covered with flattened hairs. The rostral horns divide at the tip. There is a row of hooked hairs along the outer margin of each horn .... *Pisa* ... 54

- Carapace not covered in hairs, although spines are present 55

54. Rostral horns separated by a V-shaped cleft. Moveable finger (dactylus) of the chela about one third of the total length of the propodus i.e. including the fixed finger. There is a prominent, outwardly directed, tooth on each side of the carapace towards the posterior margin giving a triangular effect. The other marginal teeth are much smaller. Carapace up to about 55 mm long. Not found on the shore .... *Pisa armata*

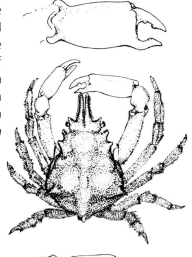

- Rostral horns separated by a Y-shaped cleft. Moveable finger (dactylus) of the chela about the total length of the propodus. The lateral borders of the carapace bear a number of sharp spines, directed forwards. Carapace up to about 50 mm long. May be found on the shore ....
.... *Pisa tetraodon*

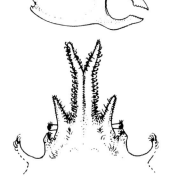

55. Rostral horns diverging. Carapace covered in spines and tubercles. Dactylus of each walking leg with a black tip. Pink in colour. Much the largest spider crab, carapace up to about 200 mm long. May be found on southwestern shores in summer ....
.... **The Spiny Spider Crab** ... *Maja squinado*

[A deep-water species *Rochinia carpenteri,* with very long rostral horns and only a few spines on the carapace, would key out here.]

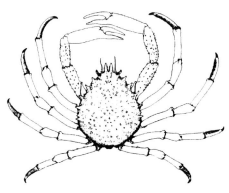

- Rostral horns converging. Carapace without spines. Dactylus of the walking legs the same colour as the rest of the leg .... *Hyas* ... 56

56. Carapace pear shaped, from the outer edge of the orbit the sides are fairly straight. Postorbital tooth small, much shorter than rostrum. Carapace about 100 mm long. May be found on the shore in the north .... *Hyas araneus*

H. araneus   *H. coarctatus*

- Carapace lyre-shaped, or "waisted". There is a decided kink in the side of the carapace. Post-orbital tooth nearly as long as the rostrum. Carapace up to about 60 mm long. May be found on the shore in the north .... *Hyas coarctatus*

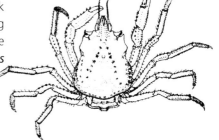

A taxonomic list of the crabs and crab-like animals of the British waters which can be identified by the use of this key

I am indebted to Dr. R. W. Ingle for bringing this list up to date. The system of classification employed here is based on that of Glaessner (1969).

Phylum: Arthropoda
  Class: Crustacea
    Sub-class: Malacostraca
      Super-order: Eucarida
        Order: Decapoda
          Division: Pleocyemata

            Sub-order: Anomura
              Super-family: Galatheoidea
                Family: Porcellanidae
                  Genus: *Porcellana* Lamarck, 1801
                    Porcellana platycheles (Pennant, 1777)
                  Genus: *Pisidia* Leach, 1820
                    Pisidia longicornis (Linneaus, 1767)   Syn: *Porcellana longicornis* (Pennant, 1777)

              Super-family: Paguroidea
                Family: Lithodidae
                  Genus: *Lithodes* Latreille, 1805
                    Lithodes maia Linneaus, 1758

                Family: Paguridae
                  Genus: *Diogenes* Dana, 1852
                    Diogenes pugilator (Roux, 1828)
                  Genus: *Pagurus* Fabricius, 1798   Syn: *Eupagurus* Brandt, 1851
                    Pagurus bernhardus (Linneaus, 1758)
                    Pagurus cuanensis Thompson, 1844
                    Pagurus prideaux Leach, 1815   Syn: *Pagurus prideauxi*
                    Pagurus pubescens Krøyer, 1838
                  Genus: *Anapagurus* Henderson, 1888
                    Anapagurus chiroacanthus Lilljeborg, 1856
                    Anapagurus hyndmanni (Thompson, 1844)
                    Anapagurus laevis (Thompson, 1844)

            Sub-order: Brachyura
              Super-family: Dromioidea
                Family: Dromiidae
                  Genus: *Dromia* Weber, 1795
                    Dromia personata (Linneaus, 1758)   Syn: *Dromia vulgaris* Milne-Edwards, 1837

Family: Corystidae
  Genus: *Corystes* Bosc, 1802
    Corystes cassivelaunus (Pennant, 1777)

Family: Portunidae
  Genus: *Portumnus* Leach, 1814
    Portumnus latipes (Pennant, 1777)   Syn: *Portumnus variegatus*
                                           Leach, 1815

  Genus: *Carcinus* Leach, 1814
    Carcinus maenas (Linneaus, 1758)   Syn: *Carcinides* Rathbun, 1897
  Genus: *Macropipus* Prestandrea, 1833   Syn: *Portunus* Fabricius, 1798
                                           in part

    Macropipus tuberculatus (Roux, 1830)
  Genus: *Liocarcinus* Stimpson, 1870   Syn: *Portunus* Fabricius, 1798
                                           in part
                                       Syn: *Macropipus* Prestandrea,
                                         1833 in part

    Liocarcinus arcuatus (Leach, 1814)
    Liocarcinus corrugatus (Pennant, 1777)
    Liocarcinus depurator (Linneaus, 1758)
    Liocarcinus holsatus (Fabricius, 1798)
    Liocarcinus marmoreus (Leach, 1814)
    Liocarcinus pusillus (Leach, 1815)
  Genus *Necora* Holthuis, 1987
    Necora puber (Linneaus, 1767)   Syn: *Liocarcinus puber*
                                         (Linneaus, 1767)

  Genus: *Polybius* Leach, 1820
    Polybius henslowii Leach, 1820   Syn: *Polybius henslowi* spelling
                                         mistake

Family: Pirimelidae
  Genus: Pirimela Leach, 1816   Syn: *Perimela* spelling mistake
    Pirimela denticulata (Montagu, 1808)

Family: Atelecyclidae
  Genus: Atelecyclus Leach, 1814
  Atelecyclus rotundatus (Olivi, 1792) nec Risso
                                         Syn: *Atelecyclus*
                                         septemdentatus (Montagu, 1808)
                                         Syn: *Atelecyclus heterodon*
                                         Leach, 1814
Family: Thiidae
  Genus: *Thia* Leach, 1815
    Thia scutellata (Fabricius, 1793)   Syn: *Thia polita* Leach, 1815

Family: Cancridae
   Genus: *Cancer* Linneaus, 1758
     Cancer pagurus Linneaus, 1758
Family: Xanthidae
   Genus: *Pilumnus* Leach, 1815
     Pilumnus hirtellus (Linneaus, 1761)
   Genus: *Monodaeus* Guinot, 1967   Syn: *Medaeus* Dana, 1851
     Monodaeus couchi (Couch, 1851)   Syn: *Xantho couchi* Bell, 1851
                                    Syn: *Medaeus couchi* (Bell, 1851)

Family: Atelecyclidae
   Genus: *Atelecyclus* Leach, 1814
     Atelecyclus rotundatus (Olivi, 1792) nec Risso
   Genus: *Xantho* Leach, 1814
     Xantho incisus Leach, 1814   Syn: *Xantho floridus*
                                      (Montagu, 1808)

     Xantho pilipes Milne-Edwards, 1867   Syn: *Xantho hydrophilus*
                                      Norman and Scott, 1906

Family: Goneplacidae
   Genus: *Goneplax* Leach, 1814   Syn: *Gonoplax* Leach, 1816
     Goneplax rhomboides (Linneaus, 1758) Syn: *Goneplax angulata*
                                      (Pennant, 1777)
Family: Pinnotheridae
   Genus: *Pinnotheres* Bosc, 1802
     Pinnotheres pinnotheres (Linneaus, 1758) Syn: *Pinnotheres veterum* Bosc, 1802
     Pinnotheres pisum (Linneaus, 1767)

Family: Leucosiidae
   Genus: *Ebalia* Leach, 1817
     Ebalia cranchii Leach, 1817   Syn: *Ebalia cranchi* spelling mistake
     Ebalia tuberosa (Pennant, 1777)
     Ebalia tumefacta (Montagu, 1808)

Family: Majidae
   Genus: *Maja* Lamarck, 1801   Syn: *Maia* spelling mistake
     Maja squinado (Herbst, 1788)
   Genus: *Pisa* Leach, 1814
     Pisa armata (Latreille, 1803)   Syn: *Pisa gibbsi* Leach, 1815
     Pisa tetraodon (Pennant, 1777)
   Genus: *Hyas* Leach, 1814
     Hyas araneus (Linneaus, 1758)
     Hyas coarctatus Leach, 1815
   Genus: *Eurynome* Leach, 1814
     Eurynome aspera (Pennant, 1777)
     Eurynome spinosa Hailstone, 1835

Genus: *Inachus* Weber, 1795
  Inachus dorsettensis (Pennant, 1777)
  Inachus phalangium (Fabricius, 1775)   Syn: *Inachus dorynchus* Leach, 1814
  Inachus leptochirus Leach, 1817
Genus: *Achaeus* Leach, 1817
  Achaeus cranchii Leach, 1817   Syn: *Achaeus cranchi* spelling
        mistake

Genus: *Macropodia* Leach, 1814   Syn: *Stenorhynchus*
  Macropodia deflexa Forest, 1978   Syn: *Macropodia aegyptia* in part
  Macropodia linaresi Forest and
   Zariquiey, 1964
  Macropodia rostrata (Linneaus, 1761)   Syn: *Macropodia phalangium*
  Macropodia tenuirostris (Leach, 1814)   Syn: *Macropodia longirostris*
        (Fabricius, 1775)

Genus: *Planes* Bowdich, 1825
  Planes minutus (Linneaus, 1758)

References

ALLEN, J. A. (1967). *The fauna of the Clyde sea area. Crustacea: Euphausiacea and Decapoda.* Scottish Marine Biological Association, Millport.

BELL, T. (1853). *A history of the British Stalk-Eyed Crustacea.* van Voorst, London.

BOSC, L. A. (1802). *Des Crustaces* in Buffon, G. L. L. Histoire Naturelle de Buffon, clasée d'après le système de Linné. Vol. 1, Paris.

BOUVIER, E. L. (1940). *Decapodes Marcheurs.* Faune de France, 37. Paris.

CHRISTIANSEN, MARIT E. (1969). *Crustacea Decapoda Brachyura.* Marine Invertebrates of Scandinavia 2, Universitetsforlaget, Oslo.

COUCH, R. Q. (1851). Notice of a crustacean new to Cornwall. *Transactions of the Natural History and Antiquarian Society of Penzance.* **2**: 13-14.

CROTHERS, J. H. (1967). The biology of the shore crab *Carcinus maenas* (L.) 1. The background – anatomy, growth and life history. *Field Studies.* **2**: 407-434.

CROTHERS, J. H. (1968). The biology of the shore crab *Carcinus maenas* (L.) 2. The life of the adult crab. *Field Studies.* **2**: 579-614.

DANA, J. D. (1851). On the classification of the Cancroidea. *American Journal of Science and the Arts.* Series 2. **12**: 121-131.

ERWIN, DAVID AND PICTON, BERNARD (1981). *The Marine Conservation Society Guide to Inshore Marine Life.* Immel Publishing, London.

FABRRICIUS, J. C. (1775). *Systema Entomologiae.*

FABRICIUS, J. C. (1798). *Supplementum Entomologiae Systematicae.*

FOREST, J. (1978). Le genre *Macropodia* Leach clans les eaux Atlantiques Européans (Crustacea Brachyura Majidae). *Cahiers de Biologie Marine,* **19**: 323-342.

FOREST, J. and ZARIEQUEY, R. (1964). Le genre *Macropodia* Leach en Méditerranée. 1. Description et étude comparative des espèces (Crustacea Brachyura Majidae). *Bulletin Museum nationale histoire naturelle, Paris.* (2) **36**: 222-224.

GLAESSNER, M. F. (1969). *Treatise on Invertebrate Palaeontology Pt. R.* Arthropoda 4pp 399-400. Geological Society of America Inc and University of Kansas.

GURNOT, D. (1967). Recherches préliminaires sur les groupements naturels chez les Crustacés Decapodes Brachyoures. 2. Les anciens genres *Micropanope* Stimpson et *Medaeus* Dans. *Bulletin Museum nationale histoire naturelle, Paris.* **39**: 345-374.

HAILSTONE, S. (1835). Descriptions of some species of crustaceous animals. *Magazine of Natural History*, VIII.

HARTNOLL, R. G. (1961). A re-examination of the spider crab *Eurynome* Leach from British waters. *Crustaceana.* **2**: 171-182.

HARTNOLL, R. G. (1963). The biology of Manx spider crabs. *Proceedings of the Zoological Society of London.* **141**: 423-496.

HENDERSON, J. R. (1888). Report on the Anomura. *Challenger Reports, Zoology* XXVII.

HERBST, J. F. W. (1792-1804). Vers. Naturg. Krabben Krebse. 3 Vols. Zurich, Berlin and Stralsund.

INGLE, R. W. (1980). *British Crabs.* British Museum (Natural History) and O.U.P.

INGLE, R. W. (1983). *Shallow water crabs. Keys and notes for the identification of the species.* Cambridge University Press.

INGLE, R. W. (1985) N.E. Atlantic and Mediterranean hermit crabs (Crustacea: Anomura: Paguridea: Paguridae) 1. The genus *Pagurus. Journal of Natural History.* **19**: 745-769.

KRØYER, H. (1838). Conspectus Crustaceorum Groenlandiae. *Naturhistorisk Tidsskrift Kjobenhavn.* **2**: 249-261.

LAMARCK, J. B. (1801). *Système des animaux sans vertèbres.*

LAMARCK, J. B. (1818). *Histoire naturelle des animaux sans vertebrès* Vol. V.

LANCASTER, I. (1988). *Pagurus bernhardus* (L.) – an introduction to the natural history of hermit crabs. *Field Studies*. **7**: 189-238.

LATREILLE, P.A. (1803). *Histoire naturelle générale et particulière des Crustacés et des Insects*. 5 Paris.

LEACH, W. E. (1813). *Crustacea* in Brewster, Ed., Edinburgh Encyclopaedia. Vol. 7: 383-437.

LEACH, W. E. (1814-1875). *Malacostraca Podophthalma Britanniae*. Sowerby, London.

LEACH, W. E. (1820). *Galatéadées*. Dictionarie des Sciences Naturele. Vol. 18: 49-56.

LILLJEBORG, W. (1851). Norges Crustaceer. *Ofv. Vet. Ak*. Forhand Stockholm. 1852.

MANUEL, R. L. (1981). *British Anthozoa*. Linnean Society Synopsis of the British Fauna No. 18. Academic Press.

MILNE-EDWARDS (1837). Histoire naturelle des Crustacés Vol. 2.

MILNE-EDWARDS, A. (1867). Description des quelques especes nouvelles de Crustacés Brachyoures. *Annales de la société entomologique de France* Paris. **7**: 263-288.

MONTAGU, G. (1808). Description of several marine animals found on the south coast of Devonshire. *Transactions of the Linnean Society of London* IX.

NORMAN, A. M. and SCOTT, T. (1906) *The Crustacea of Devon and Cornwall*. London.

PALMER, R. (1927). A revision of the genus *Portunus*. *Journal of the Marine Biological Association of the UK*. **14**: 877-908.

PENNANT, T. (1777). *British Zoology*. Vol. IV *Crustacea, Mollusca, Testacea*. London.

RATHBUN, M. J. (1897). A revision of the nomenclature of the Brachyura. *Proceedings of the Biological Society of Washington,* II, 164.

ROUX, J. S. F. P. (1828). *Crustacés de la Méditerranée et de son littoral*. Paris.
SAMOUELLE, G. (1819). *The entomologist's useful compendium*. London.

SELBIE, C. M. (1921). The Decapoda reptantia of the coasts of Ireland. Part II: Paguridea. *Scientific Investigations of the Fisheries Board for Ireland*, 20-68.

STIMPSON, W. (1870). Preliminary report of the Crustacea dredged in the Gulf Stream in the Straits of Florida. Part 1. Brachyura. *Bulletin of the Museum of Comparative Zoology, Harvard*. 2: 109-160.

THOMPSON, W. (1844). Report on the fauna of Ireland. *Report of the British Association for the Advancement of Science*, 1843.

WEBER, F. (1795). *Nomenclator entomologicus secundum entomologiam systematicum* Chilonii and Hamburgi.

Figure 3. *Lithodes maia*: Drawn from a specimen in the collection of the British Museum (Natural History). Carapace breadth 95 mm (including spines).

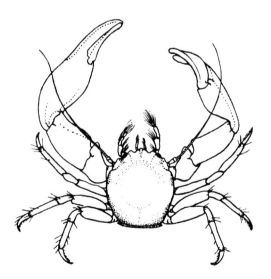

Figure 4. The Long-clawed Porcelain Crab (*Pisidia longicomis*) Male, collected from Porlock Weir, Somerset. Carapace breadth 7 mm.

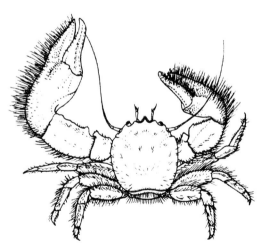

Figure 5. The Broad-clawed Porcelain Crab (*Porcellana platycheles*) Male, collected from Porlock Weir, Somerset. Carapace breadth 12 mm.

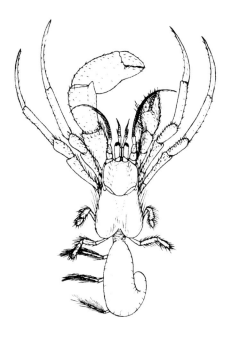

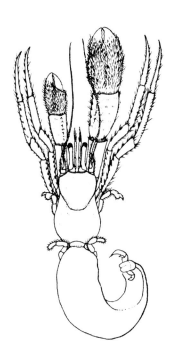

Figure 6. *Diogenes pugilator*: Drawn from a specimen in the collection of the British Museum (Natural History). Length of the propodus on the large (left) chela = 6 mm.

Figure 7. *Pagurus cuanensis*: Drawn from a specimen in the British Museum (Natural History). Length of the propodus on the large (right) chela = 7 mm.

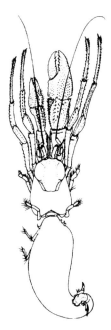

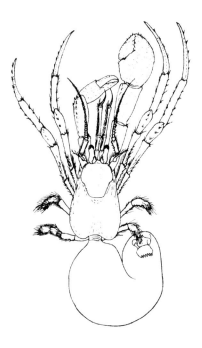

Figure 8. *Pagurus pubescens*: Drawn from a specimen in the British Museum (Natural History). Length of the propodus on the large (right) chela = 13 mm.

Figure 9. *Anapagurus hyndmanni*: Drawn from a specimen in the British Museum (Natural History). Length of the propodus on the large (right) chela = 4 mm.

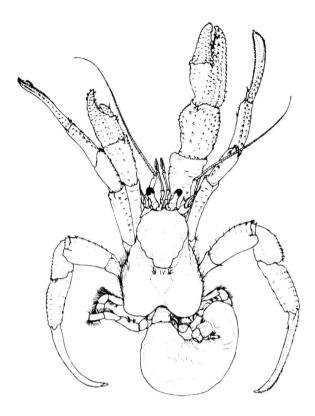

Figure 10. *Pagurus bernhardus*: A large specimen taken in Dale Roads (Milford Haven). Length of the propodus on the large (right) chela = 26 mm.

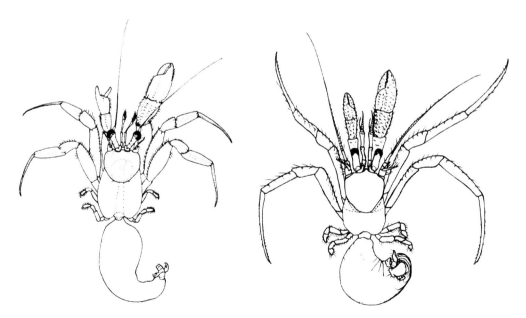

Figure 11. *Pagurus prideaux*: Drawn from a specimen collected in Milford Haven. Length of the propodus on the large (right) chela = 18.6 mm.

Figure 12. *Anapagurus laevis*: Drawn from a specimen in the British Museum (Natural History). Propodus of the large (right) chela = 4 mm.

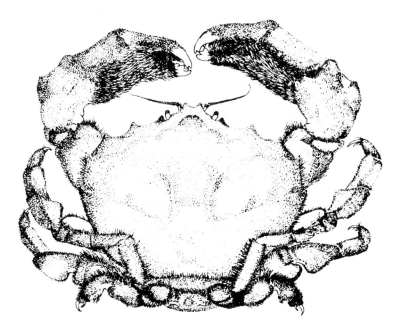

Figure 13. The Sponge Crab (*Drornia personata*): Drawn trom a specimen in the British Museum (Natural History), Carapace 28.5 mm across.

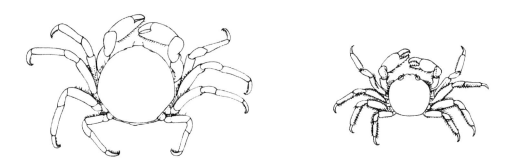

Figure 14. The pea crab *Pinnotheres pisum*: female, 8 mm across the carapace, on the left. Male, 4 mm across, on the right. Drawn from specimens in the British Museum (Natural History).

Figure 15. The pea crab *Pinnotheres pinnotheres*: female, 8 mm across the carapace, on the left. Male, 4 mm across, on the right. Drawn from specimens in the British Museum (Natural History).

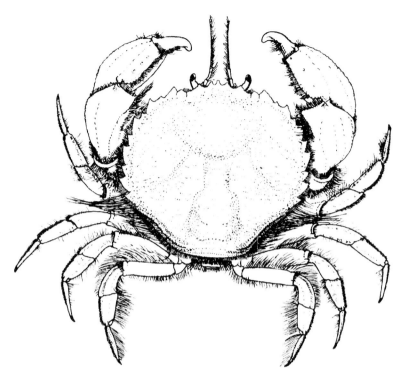

Figure 16. The Circular Crab *Atelecyclus rotundatus*. Drawn from a specimen collected in Milford Haven. Carapace 50 mm across.

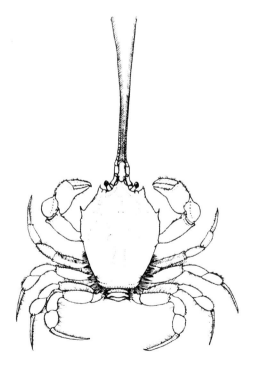

Figure 17. The Masked Crab (*Corystes cassivelaunus*): Drawn from a female collected in Milford Haven. Carapace 57 mm across.

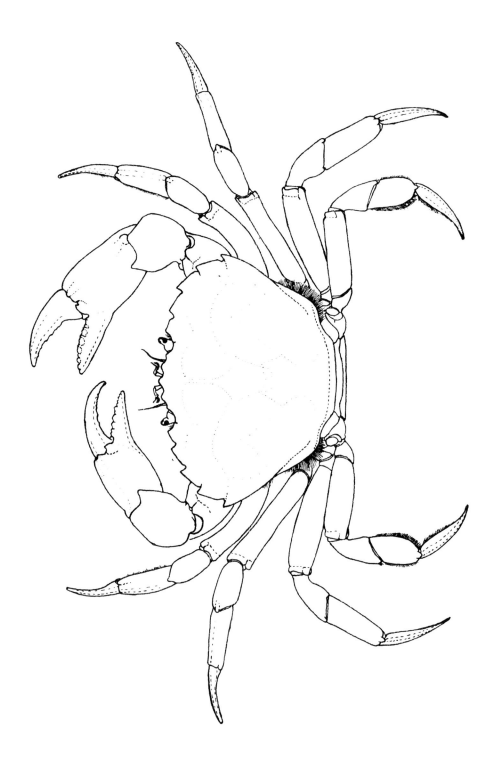

Figure 18. The Shore Crab (*Carcinus maenas*): a large male collected in Sullom Voe (Shetland). Carapace 88 mm across.

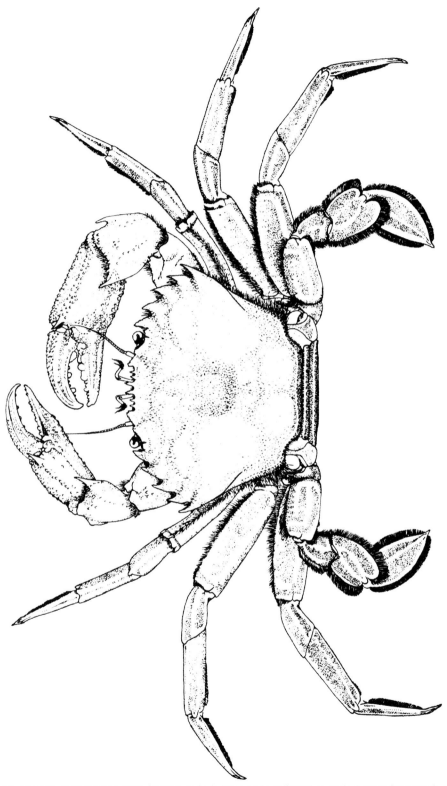

Figure 19. The Velvet Swimming Crab (*Necora puber*): large male collected at Porlock Weir (Somerset). Carapace 96 mm across.

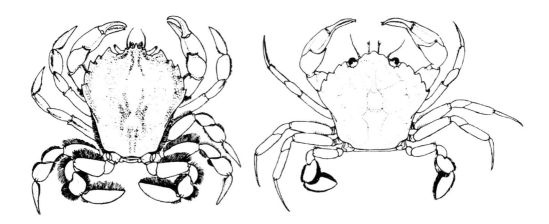

Figure 20. *Portumnus latipes*: Drawn from a specimen collected on Marloes Sands, Dyfed. Carapace 25 mm across.

Figure 21. *Liocarcinus pusillus*: male. Drawn from a specimen collected in Dale Roads (Milford Haven). Carapace 25 mm across.

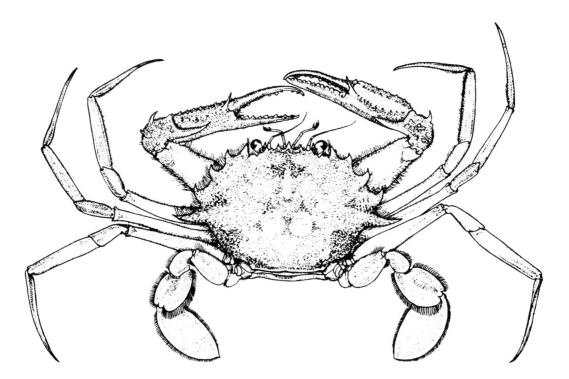

Figure 22. *Macropipus tuberculatus*: male. Drawn from a specimen in the British Museum (Natural History). Carapace 75 mm across.

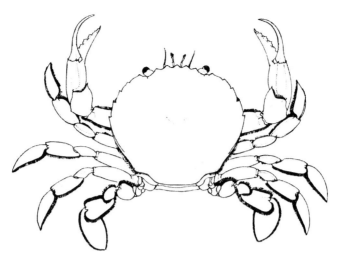

Figure 23. *Polybius henslowii*: Drawn from a male specimen collected in Milford Haven. Carapace 63 mm across.

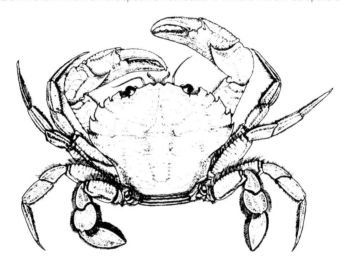

Figure 24. *Liocarcinus corrugatus*: male. Drawn from a specimen in the British Museum (Natural History). Carapace 100 mm across.

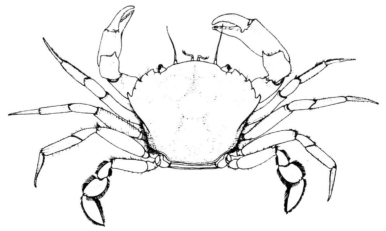

Figure 25. *Liocarcinus arcuatus*: Drawn from a male specimen trapped in Dale Roads, Milford Haven. Carapace 39 mm across.

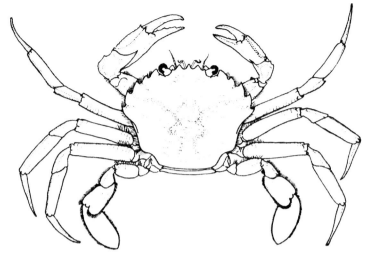

Figure 26. *Liocarcinus depurator*: male. Drawn from a specimen trapped in Dale Roads (Milford Haven). Carapace 58 mm across. The specimen is regenerating the right chela, which should be larger than the left.

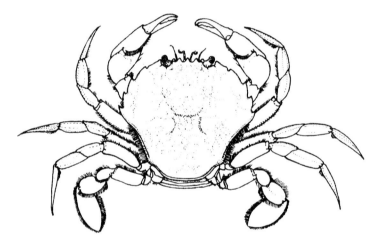

Figure 27. *Liocarcinus marmoreus*: female. Drawn from a specimen collected on Handa, Inner Hebrides.

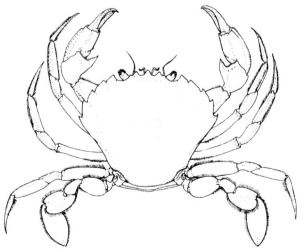

Figure 28. *Liocarcinus holsatus*: male. Drawn from a specimen in the British Museum (Natural History).

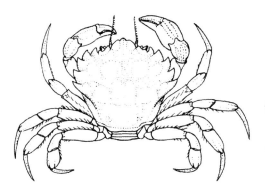

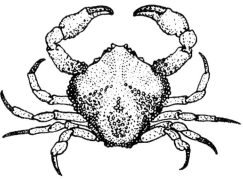

Figure 29. *Pirimela denticulata*: Drawn from a specimen in the British Museum (Natural History). Carapace 12 mm across.

Figure 30. *Ebalia tuberosa*: male. Drawn from a specimen in the British Museum (Natural History).

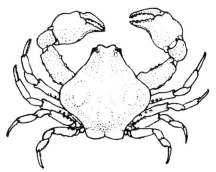

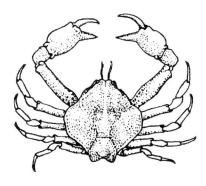

Figure 31. *Ebalia tumefacta*: male. Drawn from specimens in the British Museum (Natural History).

Figure 32. *Ebalia cranchii:* male. Drawn from a specimen in the British Museum (Natural History).

Figure 33. *Thia scutellata*. Drawn from a specimen in the British Museum (Natural History).

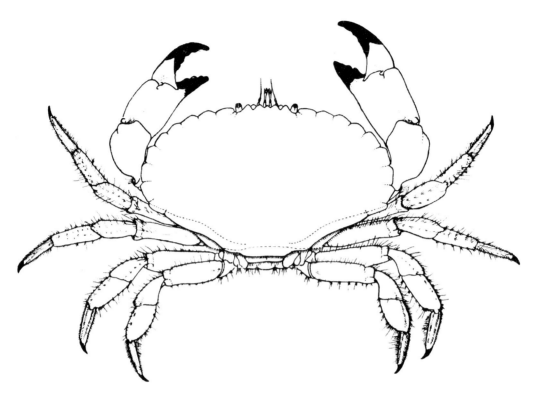

Figure 34. The Edible Crab (*Cancer pagurus*): Drawn from a specimen trapped in Dale Roads (Milford Haven). Carapace 130 mm across.

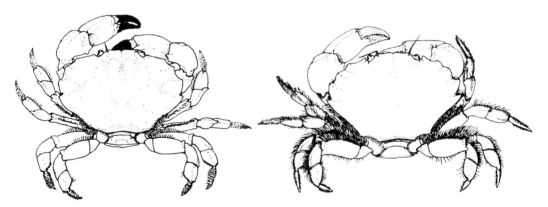

Figure 35. *Xantho incisus*: female. Drawn from a specimen collected in Milford Haven. Carapace 32 mm across.

Figure 36. *Xantho pilipes*: female. Drawn from a specimen collected at Porlock Weir (Somerset). Carapace 28 mm across.

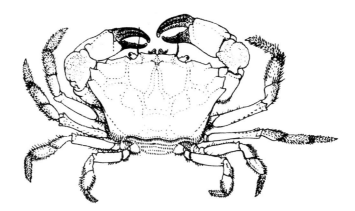

Figure 37. *Monodaeus couchi*: male. Drawn from a specimen in the British Museum (Natural History). Carapace 32 mm across.

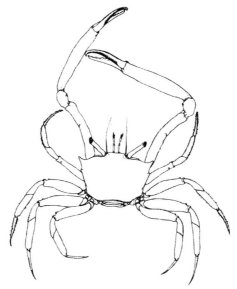

Figure 38. *Goneplax rhomboides*: male. Drawn from a specimen collected in Dale Roads (Milford Haven). Carapace 45 mm across.

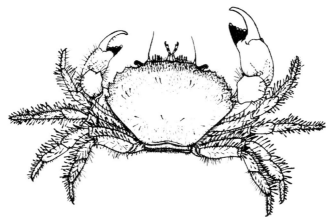

Figure 39. The Hairy Crab (*Pilumnus hirtellus):* male. Drawn from a specimen collected at Porlock Weir (Somerset). Carapace 32 mm across.

Figure 40. *Inachus dorsettensis*: male. Drawn from a specimen in the British Museum (Natural History). The specimen has both chelae tucked under the body so the shape is indicated by the inset. Carapace 24 mm across the base.

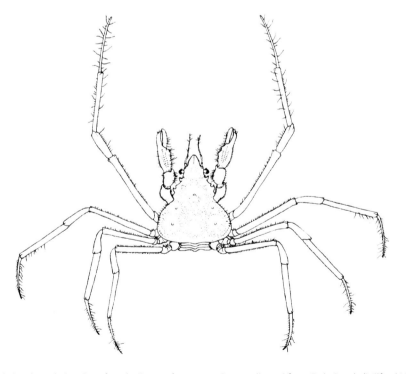

Figure 41. *Inachus phalangium*: female. Drawn from a specimen collected from Dale Roads (Milford Haven). Carapace 15 mm across the base. *I. leptochirus* is similar apart from the rostral horns and eyestalks.

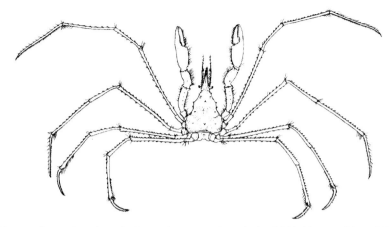

Figure 42. *Macropodia tenuirostris*: male. Drawn from a specimen in the British Museum (Natural History). Carapace 15 mm across the base.

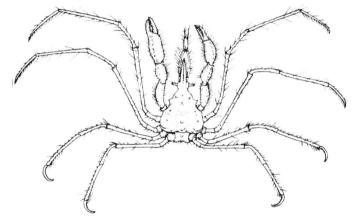

Figure 43. *Macropodia deflexa*: male. Drawn from a specimen in the British Museum (Natural History). Carapace 12 mm across the base.

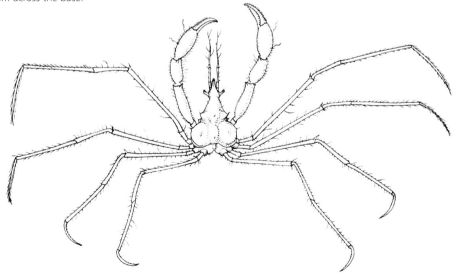

Figure 44. *Macropodia rostrata*: male. Drawn from a specimen collected in Dale Roads (Milford Haven). Carapace 15 mm across the base.

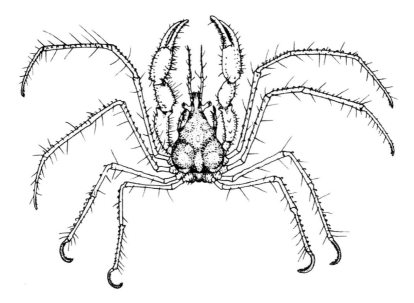

Figure 45. *Macropodia linaresi*: male. Drawn from a specimen in the British Museum (Natural History). Carapace 8 mm across the base.

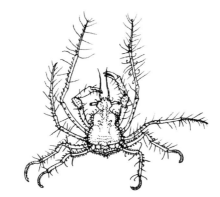

Figure 46. *Achaeus cranchii*: Drawn from a specimen in the British Museum (Natural History). Carapace 5 mm across.

Figure 47. *Eurynome aspera*: female. Drawn from a specimen in the British Museum (Natural History). Carapace 12.5 mm across.

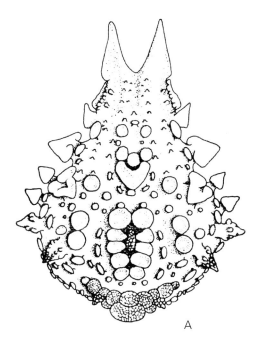

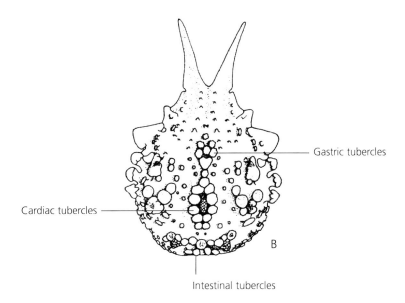

Figure 48. Dorsal surfaces of carapace: (A) *Eurynome aspera*, (B) *E. spinosa*. Reproduced from Hartnoll (1963) Fig. 1 by permission of the author and The Zoological Society of London.

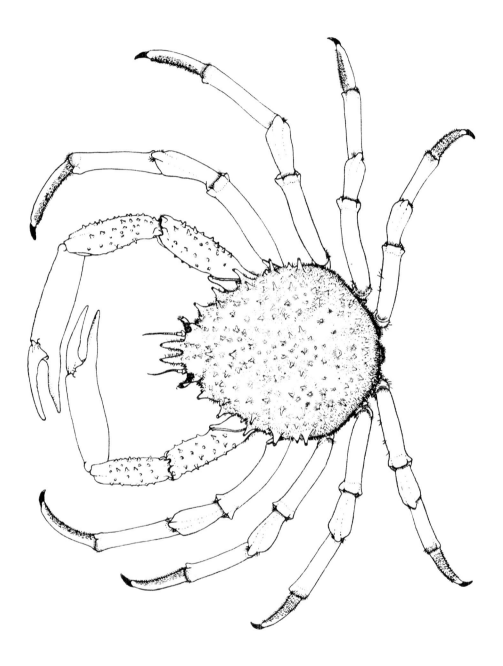

Figure 49. The Spiny Spider Crab (*Maja squinado*): male. Drawn from a specimen collected in Milford Haven. 203 mm across the carapace.

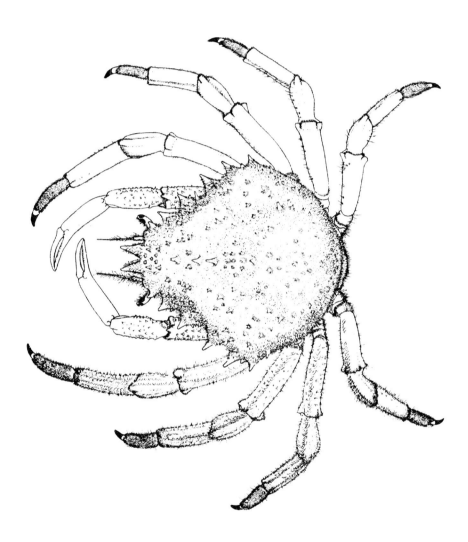

Figure 50. The Spiny Spider Crab (*Maja squinado*): female. Drawn from specimens collected in Milford Haven. Left side shows the legs of a newly-moulted crab, the right side shows the legs after the hairs have been worn away. Carapace 150 mm across.

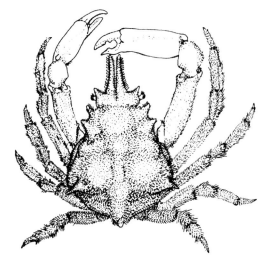

Figure 51. *Pisa armata*: male. Drawn from a specimen in the British Museum (Natural History). Carapace 30 mm across.

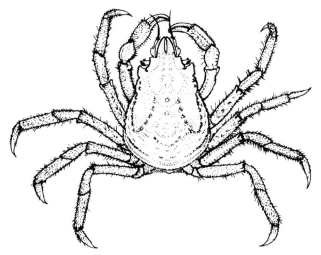

Figure 52. *Hyas araneus*: male. Drawn from a specimen in the British Museum (Natural History). Carapace 31 mm across.

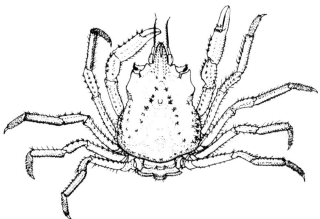

Figure 53. *Hyas coarctatus*. Drawn from a specimen in the British Museum (Natural History). Carapace 25 mm across.

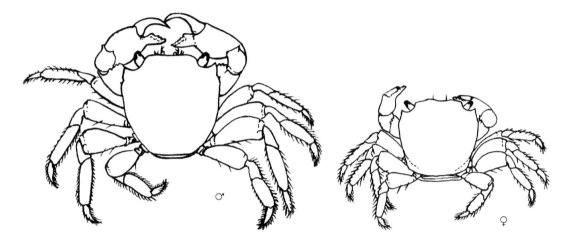

Figure 54. *Planes minutus*. Male (left) 12.7 mm and female (right) 13.0 mm carapace breadth. Drawn from specimens collected off Fair Isle (59'33'N:1°38'W) by the crew of the "Good Shepherd" 28 November 1989.